My First Science Books
A CRABTREE SEEDLINGS BOOK

I see Light

Francis Spencer

CRABTREE
PUBLISHING COMPANY
WWW.CRABTREEBOOKS.COM

Look around! You are using your eyes to see.

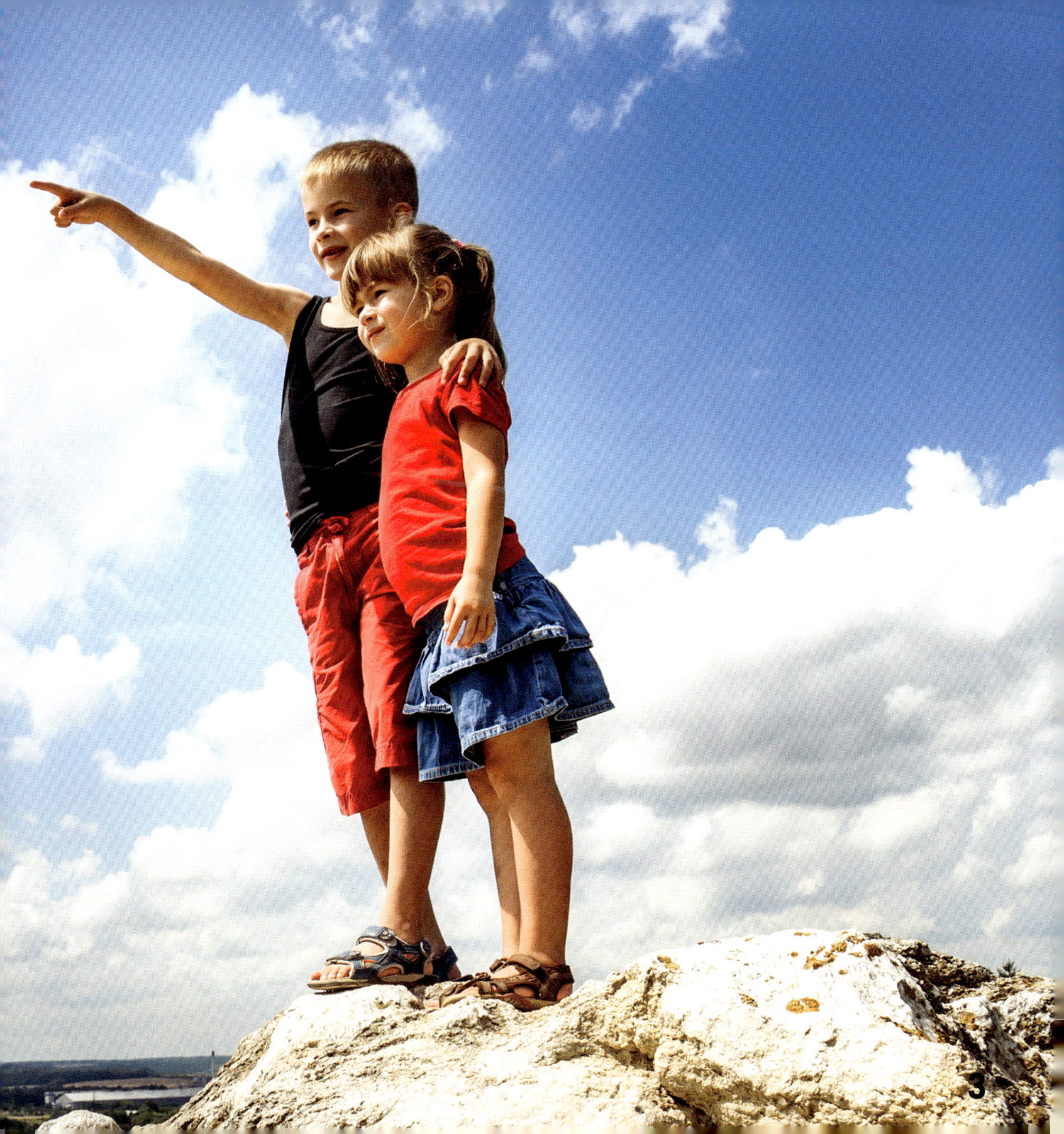

Seeing is one of our five **senses**.

We use our senses to learn about our world.

The 5 Senses:

seeing

touching

hearing

smelling

tasting

We have two eyes for seeing.

The colored part of your eye is called the iris (EYE-riss).

Many Animals Have More Than Two Eyes!

A praying mantis has five eyes.

A scorpion has 10 to 12 eyes.

A scallop can have up to 200 eyes!

We need **light** to see. Without light, everything would be in darkness.

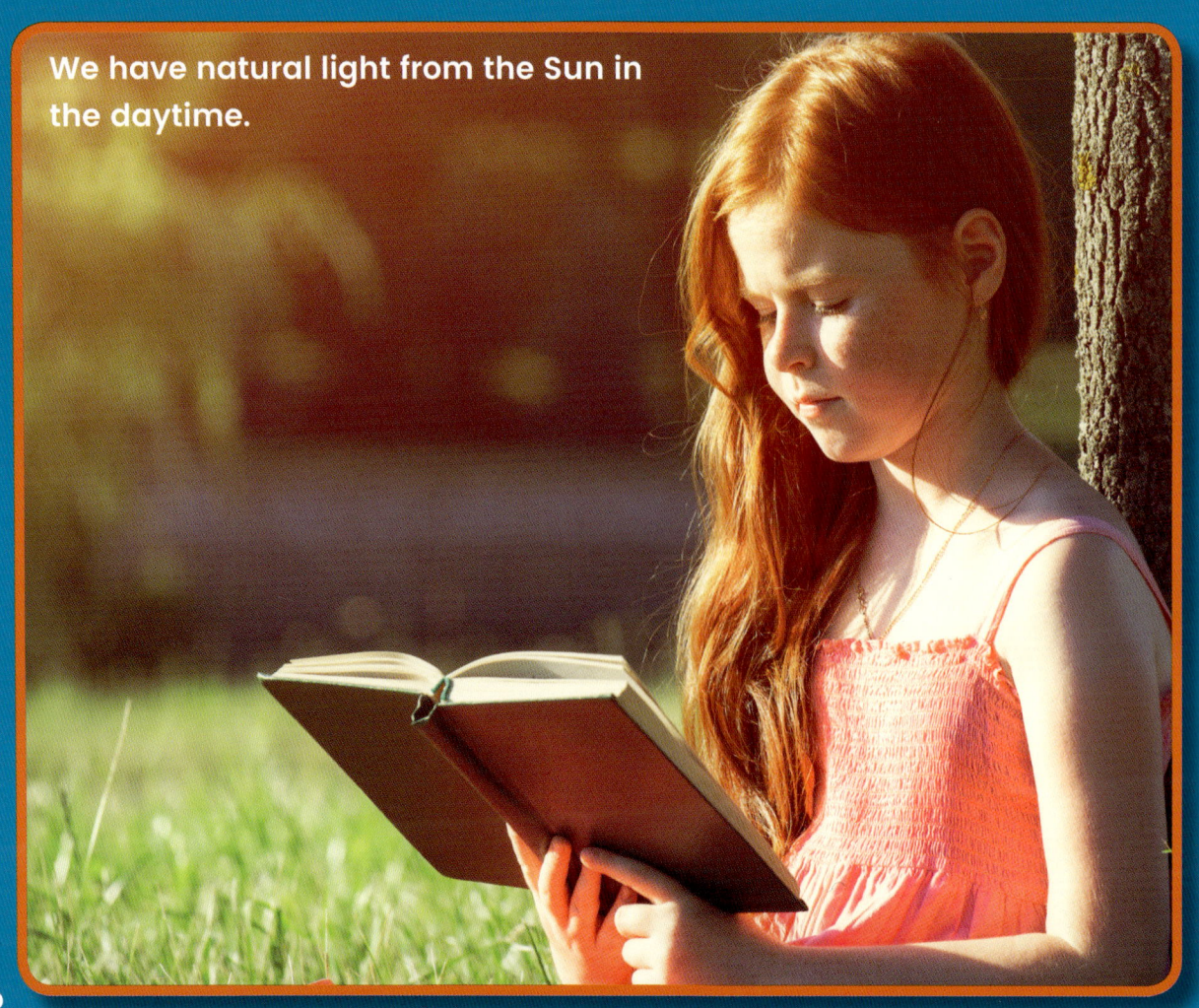

We have natural light from the Sun in the daytime.

When it's dark, we use flashlights and lamps to make light.

Light passes through some objects. It does not pass through others.

Light can shine through anything that is **transparent**. Glass, water, air, and the lens in your eye are transparent.

10

Translucent objects let only some light shine through. Tracing paper, sunglasses, and thin cloth are translucent.

Opaque objects do not let any light shine through. Wood, metal, and clay are opaque.

11

Light shines on objects. It then bounces back and goes into our eyes through our pupils.

pupil
(PYOO-puhl)

The pupil is a hole that lets light into your eye.

13

After light goes into the pupil, it shines through the lens. The lens is transparent. The lens helps you see things in **focus**.

Out of focus

In focus

14

retina
(RET-in-uh)

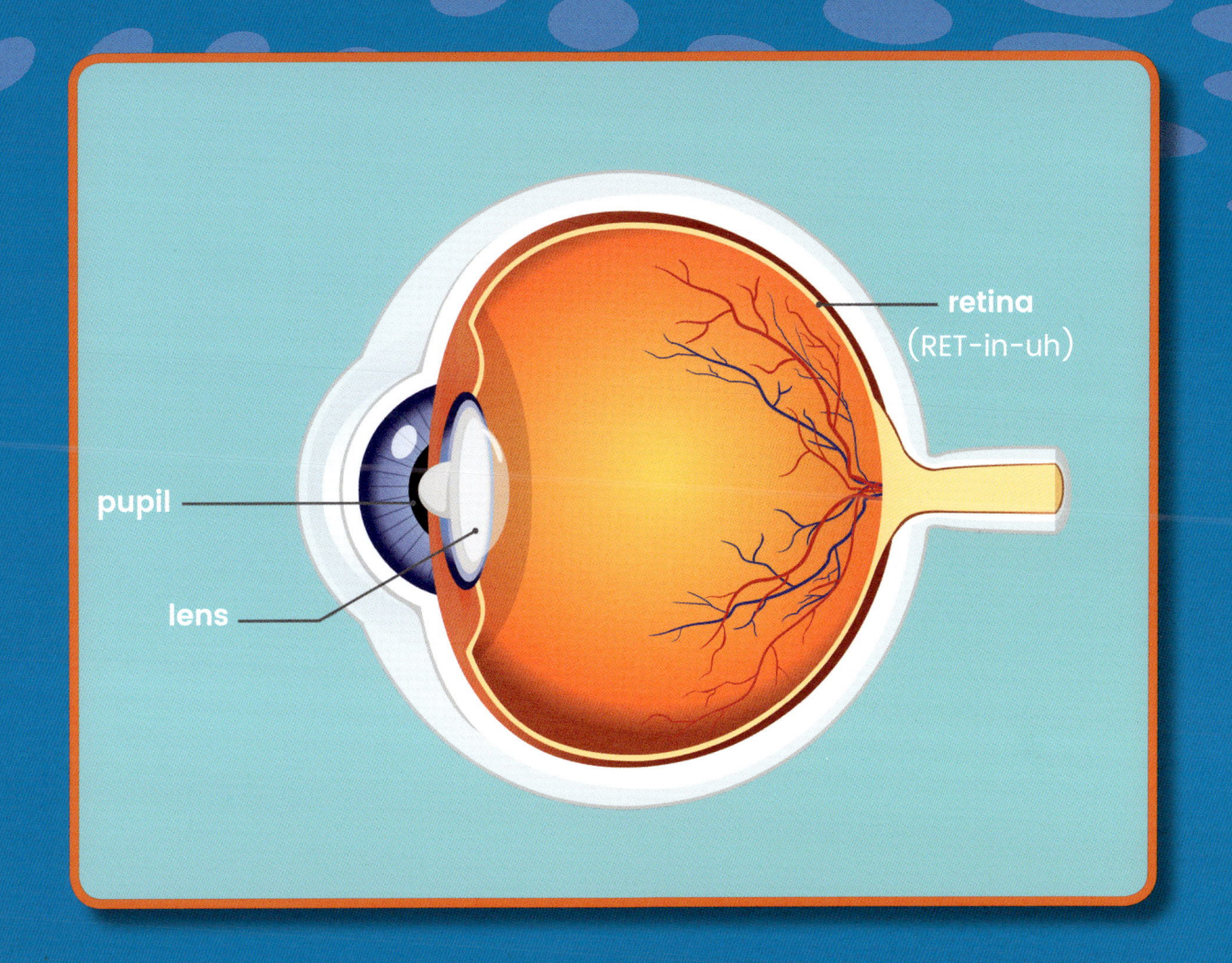

pupil

lens

Light hits the retina after going through the lens. The retina sends messages to the brain.

15

Your brain uses the messages to make pictures.

The pictures tell you the shape, size, and color of everything you see:

a big, red ball

a small, green frog

17

Many animals have an excellent sense of sight. It helps them to hunt and survive.

Eagles have super powerful eyesight. They can see a rabbit 2 miles (3.2 kilometers) away.

A chameleon's eyes can look in two different directions at the same time!

18

Our pupils look like circles, but a goat's pupils look like rectangles. The long shape means that goats can see forward and side-to-side at the same time.

A tarsier has huge eyes that allow it to see perfectly in the dark!

Helpful Tools

There are tools we can use to help us see better.

Glasses are lenses that help people who cannot see well.

Binoculars make faraway things look closer.

A hand lens makes small things look bigger.

Telescopes give us a better look at stars, galaxies, and planets.

What Did You Learn?

Read each question, then choose an answer. Find a sentence in the book that proves your answer.

1. Which part of your eye sends messages to your brain?

    pupil   lens   retina

2. Which of the objects below is transparent?

3. What would be a useful tool for seeing in the dark?

    glasses   flashlight   binoculars

21

Glossary

In focus

focus (FOH-kuhss): When you see things in focus, you see them clearly.

light (LITE): Light is the bright energy we need to see. We get natural light from the Sun. People make light with flashlights, lamps, and candles.

opaque (oh-PAKE): Opaque materials do not let any light shine through. We cannot see through opaque materials.

senses (SENSS-ez): Senses are the five powers we use to help us enjoy and stay safe in our world: touching, seeing, tasting, smelling, hearing.

translucent (trans-LOO-suhnt): Translucent materials let only some light shine through.

transparent (trans-PAIR-uhnt): Transparent materials let light shine through. We can see through transparent materials.

Index
light 8, 9, 10, 11, 12, 13, 14, 15
opaque 11
pupil(s) 12, 13, 14, 15, 19
seeing 4, 5, 6
senses 4, 5, 18
tools 20
translucent 11
transparent 10, 14

School-to-Home Support for Caregivers and Teachers

Crabtree Seedlings books help children grow by letting them practice reading. Here are a few guiding questions to help the reader with building his or her comprehension skills. Possible answers are included.

Before Reading
- **What do I think this book is about?** I think this book is about how people see light. It might tell us about how our eyes work.
- **What do I want to learn about this topic?** I want to learn about where light comes from. I see light from candles on the cover.

During Reading
- **I wonder why...** I wonder why some animals have so many eyes.
- **What have I learned so far?** I have learned that light comes from natural sources, such as the Sun, and other sources such as flashlights and lamps.

After Reading
- **What details did I learn about this topic?** I learned that light enters our eyes through our pupils.
- **Read the book again and look for the vocabulary words.** I see the word *senses* on page 4 and the word *light* on page 8. The other vocabulary words are found on pages 22 and 23.

Library and Archives Canada Cataloguing in Publication

Title: I see light / Francis Spencer.
Names: Spencer, Francis, 1973- author.
Description: Series statement: My first science books | "A Crabtree seedlings book". | Includes index. | Previously published in electronic format by Blue Door Education in 2020.
Identifiers: Canadiana 20200389556 | ISBN 9781427130259 (hardcover) | ISBN 9781427130365 (softcover)
Subjects: LCSH: Vision—Juvenile literature. | LCSH: Eye—Juvenile literature. | LCSH: Light—Juvenile literature.
Classification: LCC QP475.7 .S64 2021 | DDC j612.8/4—dc23

Library of Congress Cataloging-in-Publication Data

Names: Spencer, Francis, 1973- author.
Title: I see light / Francis Spencer.
Description: New York, NY : Crabtree Publishing Company, 2021. | Series: My first science books ; a Crabtree seedlings book | Includes index.
Identifiers: LCCN 2020051005 | ISBN 9781427130259 (hardcover) | ISBN 9781427130365 (paperback)
Subjects: LCSH: Light--Juvenile literature.
Classification: LCC QC360 .S653 2021 | DDC 535--dc23
LC record available at https://lccn.loc.gov/2020051005

Crabtree Publishing Company
www.crabtreebooks.com 1–800–387–7650

e-book ISBN 978-1-949354-02-7
e-pub ISBN 978-1-949354-03-4

Print book version produced jointly with Blue Door Education in 2021

Author: Francis Spencer
Production coordinator and Prepress technician: Tammy McGarr
Print coordinator: Katherine Berti

Printed in the U.S.A./012021/CG20201112

Content produced and published by Blue Door Publishing LLC dba Blue Door Education, Melbourne Beach FL USA. Copyright Blue Door Publishing LLC. All rights reserved. No part of this book may be reproduced or utilized in any form or by any means, electronic or mechanical including photocopying, recording, or by any information storage and retrieval system without permission in writing from the publisher.

Photo credits: Cover © A3pfamily; page 4-5 © Bilanol; page 6-7 illustrations © Gorbachev Oleg, photo of girl © Rawpixel.com; page 8 © Gelpi, page 9 mantis © Muhammad Naaim, scorpion © IrinaK, scallop courtesy of NOAA; page 10 © A. and I. Kruk, page 11 © SeventyFour; page 12 boy © alexandre zveiger, sun © knahthra, model plane © Lyubov Timofeyeva; page 13 © Peter Kotoff; page 14 © REDPIXEL.PL ; page 15 ladybug © irin-k; page 16 © solar22, page 17 human © Vectormine, thought bubble © akiradesigns; page 18 ball © Alexander Lukyanov, frog © Kuttelvaserova Stuchelova; page 19 eagle © Colin Edwards Wildside, chameleon © Chantelle Bosch, goat © AnnaTamila; page 20 glasses © Africa Studio, binoculars MNStudio, hand lens © paulista, telescope © paulista; page 21 clear piggy bank © pogonici, blue piggy bank © g215, pink piggy bank © Andrey Burmakin; page 22 red mug © Africa Studio; page 23 child with glass © Ann in the uk, frog on leaf © Jonathan Chancasana. All images from Shutterstock.com.

Published in Canada
Crabtree Publishing
616 Welland Ave.
St. Catharines, Ontario
L2M 5V6

Published in the United States
Crabtree Publishing
347 Fifth Ave.
Suite 1402-145
New York, NY 10016

Published in the United Kingdom
Crabtree Publishing
Maritime House
Basin Road North, Hove
BN41 1WR

Published in Australia
Crabtree Publishing
Unit 3 – 5 Currumbin Court
Capalaba
QLD 4157